How to Increase IQ

Discover How to Raise Your IQ and Increase Your Intelligence

by **Remy Simon**

Table of Contents

Introduction 1

Chapter 1: Understanding IQ Testing 11

Chapter 2: How to Evaluate the Different Aspects of IQ 15

Chapter 3: Going Beyond the Test..25

Chapter 4: Lifestyle Tips to Improve IQ...................................33

Chapter 5: Using Decision Making to Enhance Intelligence 37

Conclusion..41

Introduction

Years of research and study has led to a wide variety of interrelated conclusions as to what intelligence really means, but the simplest yet over-arching definition of the word is this: that intellect is the capacity of an individual to acquire, understand, and use knowledge.

Along with the general definition above, the fact that there are many definitions to intelligence means that it can manifest in as many ways as there are people who are exercising intellect. There is therefore no single measurement or style in showing how smart a person is or can be.

To demonstrate, intelligence does not necessarily mean:

- good memory, which only refers to retention of knowledge, but not the use of it;

- self-awareness, which is somewhat more instinctive and not knowledge dependent; or

- being educated, because that only requires to go through an academic institution.

These, among many others (i.e. emotional perception, logic, etc.), are but subcategories of the broader concept of intelligence.

Intelligence is a very important factor in life because it is one of the essential elements to self-improvement. It is through the personal understanding of information and the ability to predict the outcomes of its possible applications, that one is able to complete tasks like:

- troubleshoot all sorts of problems;

- come up with new ideas to improve existing ones; and

- think of long term consequences of a certain action

Simply put, being smart is what allows us to do great things in the real world. While it may be true that it is not the only source of success (because the real world has proven that even the less intelligent are able to thrive better than others in life), it is still something that people should never take for granted.

More important, is the idea that intelligence can still be further improved. Statistics have shown that intellect is comprised of 40% genetics and 60% acquired skill. You will learn later that this sort of awareness is important because thinking otherwise can inhibit any potential growth on a person's intelligence. Take note that intelligence can be improved regardless of a person's age.

There are many ways to measure intellect, but the one that's been used since time immemorial is the intelligence quotient or IQ, which is the subject matter of this guide. Despite having been used for years, this measurement has been constantly challenged, and its conventional meaning (which will be elaborated on later) has been expanded to include what is now known as the different aspects of intelligence.

Aspects of Intelligence

So what kinds of intelligences are there? So far, science has been able to classify it according to the following:

Linguistic or Verbal. This pertains to effective use of language and vocabulary. It is exercised by debaters and public speakers.

Bodily or Kinesthetic. Here, the brain can manipulate the body to perform physical functions accurately. Talented dancers and athletes have body intellect.

Visual or Spatial. People of this type of intellect tend to perceive more than what the average eye can see. This intelligence is common among artists and designers.

Logical or Mathematic. This is manifested by strong problem-solving ability, especially when numbers and figures are involved.

Interpersonal or Social. Those with this intellect are really good at picking up social cues and adjusting to help person to person interactions work well.

Intrapersonal. In contrast to interpersonal relationships, this refers to the ability to relate with oneself. People with good intrapersonal skill are quite focused and know how to mind themselves.

Musical or Rhythmic. This is the intelligence of musicians of all kinds. They are able to understand music like it's their second language.

Creativity. This is the ability of person to think outside the box. Those who master creativity are able to think of new ideas and share them with other people.

Memory. Formerly seen as a simple brain function, it has been recognized that there are people who are able to retain facts and information much better than others.

The above categorization came from the theory of multiple intelligences, which contests the concept of general knowledge for being the only way to measure intelligence across the board. It was the idea of a general knowledge that made people think that some people just can't become more intelligent than others.

A corollary to this theory is that while everyone will possess the aforementioned intelligences in one way or another, no two persons will have the same capacity to develop each one. A person can be especially good with music, although not necessarily efficient with the other intelligences, while another can be exceptional at math and just be average with the others.

It is very possible for a person to excel at more than one aspect of intelligence. In fact, a person can even proceed to develop more, if not all of these intelligences with the right kind of discipline and training.

In view of all this, we can dispel the idea that a person is stuck with what intellect he or she is born with. Knowing which aspects of intellect you are naturally good at, as well as which ones you are not, will allow you to take steps towards self-improvement.

This guide is not just some source of IQ tests (save a few examples). This guide is going to delve deeper into the evolving concept of IQ as well as introduce tried and tested practices and techniques of improving them, explained in ways that the average person can understand.

© Copyright 2014 by Miafn LLC - All rights reserved.

This document is geared towards providing reliable information in regards to the topic and issue covered. The publication is sold with the idea that the publisher is not required to render accounting, officially permitted, or otherwise, qualified services. If advice is necessary, legal or professional, a practiced individual in the profession should be ordered.

- From a Declaration of Principles which was accepted and approved equally by a Committee of the American Bar Association and a Committee of Publishers and Associations.

In no way is it legal to reproduce, duplicate, or transmit any part of this document in either electronic means or in printed format. Recording of this publication is strictly prohibited and any storage of this document is not allowed unless with written permission from the publisher. All rights reserved.

The information provided herein is stated to be truthful and consistent, in that any liability, in terms of inattention or otherwise, by any usage or abuse of any policies, processes, or directions contained within is solely and completely the responsibility of the recipient reader. Under no circumstances will any legal responsibility or blame be held against the publisher for any reparation, damages, or monetary loss due to the information herein, either directly or indirectly.

Respective authors own all copyrights not held by the publisher.

The information herein is offered for informational purposes solely, and is universal as so. The presentation of the information is without contract or any type of guarantee assurance.

The trademarks that are used are without any consent, and the publication of the trademark is without permission or backing by the trademark owner. All trademarks and brands within this book are for clarifying purposes only and are the owned by the owners themselves, not affiliated with this document.

Chapter 1: Understanding IQ Testing

Intelligence Quotient is a very old way of measuring intellect through a standardized test. It is that standardization that has allowed this approach to effectively compare two or more people of the same class. The reason why it is called a quotient is because division is involved. That is, one specific quantity over another, which depends on the test given.

There are different kinds of IQ tests, with each one attempting to measure a specific aspect of intellect. The comparison is then made based on how well each and every one who took the same test responds to it. Because traditional IQ tests assume that people have a general ability of the mind, it is made part of an entire set of psychometric tests that aim to measure one's mind. Through a combination of tests, a person's personality, traits, and abilities are measured to an extent that can be compared with all the people who also took the same test. In turn, these are called aptitude tests.

IQ is also best tested and classified between the ages 13 and 18. This is because IQ testing advocates believe that consistent development takes place before reaching thirteen years old, from which the development begins to slow down, lowering to a minimum when reaching the age of 18. As such, IQ is measured differently between children and adults.

Standard Criteria for IQ

The first standard that is common among all IQ tests is time. Tests are often administered with a time limit because how fast a person resolves these tests is also indicative of ability.

The second standard would be the type of tests, which will be further discussed in the next chapter. A single IQ test will normally be divided into parts, comprising verbal reasoning, numerical or schematic ability, and spatial reasoning. These are then further divided into subparts.

Is IQ testing really enough?

Not in its traditional sense, it isn't. The main weakness of standardized IQ tests is that it's very easy to rehearse or even get familiar with the kind of tests that are being administered. So at some point, there is a ceiling to the effectiveness of these tests.

Also, traditional IQ tests do not measure other kinds of intelligences.

But this is not to say that one should not try to take these exams. They nonetheless cover a huge portion of evaluating intellect, so it is still important to the whole process. It's just that it cannot be the only reference if you want a full evaluation of your own intelligence, which is necessary for self-improvement.

Chapter 2: How to Evaluate the Different Aspects of IQ

A good understanding of the different parts of psychometric testing will allow you to make sure that everything is included in your checklist of tests to take. While these tests will already help improve your IQ, they are primarily designed to measure your current standing among different types of intelligences.

Note: As mentioned before, this guide will not provide a complete set of tests to work on. These tests can be better provided by experts or licensed publications. The goal of this guide is to simply help you learn what you're looking out for.

Basic Aptitude

Aptitude, although oftentimes interchanged with ability or achievement, more aptly refers to one's potential to learn new things in the future. The basics include verbal, numerical, and technical tests.

Verbal Aptitude measures your ability to perceive and convey language (in this case, the English language). High scorers are able to use language effectively in what they do. These usually involve vocabulary tests such as basic synonym and antonym tests, as well as analogy tests (i.e. context clues). Sometimes you will be tested on your reading comprehension skills, where a short anecdote is a precursor to basic questions.

An example for a verbal aptitude test would one that would require you to select which among a set of given words do not belong in that list. For instance:

Up is to down as left is to _____.

In the above example, the answer would be "right" because that is the opposite of the word left. The first part of the question assumes that the word being sought is an opposite of the first word.

Improving your verbal aptitude is perhaps one of the simplest, most straightforward things to do. All you have to do is read and practice using the words you have learned and understood. You are likely to excel in verbal aptitude tests if you are well-versed in language and vocabulary, because naturally, that is what a person with good verbal aptitude is all about.

Numerical Aptitude falls under the logical aspect of intelligence, with a focus on numbers. Yes, this is where a lot of math comes in. But unlike language tests, numbers are universal and therefore apply to all regardless of culture. Numerical tests are usually very accurate measurements of logic largely because math is absolute—it all needs a very logical mind to manipulate it.

Ranging from numerical order and relations, down to basic math problems, the types of numerical aptitude tests are endless.

0, 1, 4, 9, _____

Given the sequence above, the answer would be 16 because the pattern shows that every succeeding number is summed with the next odd number, starting from 1.

The key to getting good at numbers and math in general is really practice—both with the hand and the brain. The use of calculators are normally not allowed during testing because you are also being assessed for how well you can do the problem in your head. Generally, the better and sharper you are with numbers, the higher your numerical aptitude is.

Technical Aptitude refers to mechanical reasoning. The test aims to determine how well a person can manage technical tasks through a series of simple problems that need to be solved.

Unlike the first two aptitude test types, technical aptitude has no clear classification of its problems. The only thing they have in common is that the tests require you to solve a given problem. It ranges from determining the total number of boxes in a given stack to finding dimensions of a given object.

Much like math, technical aptitude requires practice. It does, however, require visual creativity as well, since you will need to visualize solving the technical problems in your head (these tests are always on paper and thus, hardly tactile). A bit of knowledge with physics is also an advantage.

Advanced Tests

You'll notice that the basic aptitude tests above will require a bit of logic, creativity, and some other skill. In the advanced set of tests, you will encounter tests that focus heavily on a particular niche. Naturally, they will tend to have increased difficulty.

Creativity. This isn't just about being artistic. Creativity exists on both the left and right hemispheres of the brain. It refers to a person's mental process (i.e. how he or she thinks). It is often referred to as the "eighth intelligence" because it was never originally part of the seven aspects of intellect.

Creativity can be tested in many ways. There are personality tests, which require you to rate a type of behavior or trait in a scale, based on how you see yourself. There are also exercises that require you to deal with imaginary shapes or optical illusions. There are also tests that require a bit of logic and lateral thinking. Many of these still have a given answer, while others only seek to determine how a person thinks (and there is no right or wrong answer to that).

> *Example of a creative personality test: How organized are you? Rate yourself using a scale of 1 to 5, with 1 being 'not at all organized' and 5 being 'highly organized'.*

> *Example of creative logic: In a given set of 3D objects, which doesn't belong to the group?*

Creativity can be improved. Generally, keeping an open mind so as not to dull the senses helps creativity grow. There are also certain activities and lifestyle choices that you can engage in to improve how your mind thinks, which will be discussed in the subsequent chapters.

Logic. Although similar to creativity because it measures a specific mental process, logic is more dependent on reason than imagination. An analysis will always point to the correct answer.

Logic tests will resemble numerical tests in some way, but you'll realize that the numbers are used in a different way. Sometimes the numbers will be substituted with words. From "odd one out" to "what comes next" questions, logic tests will require you to think. Needless to say, tests under this category will require more than just a calculator or formula to accomplish. Here's an example of a very complex logic question:

> *A says, "D was the one who did it". B says, "Not me". C says, "Neither did E." D says, "Not me either". E says, "B is correct." Given the facts, who is telling the truth?*

This is a simple "crime-solving" question, and there are many more.

Much like math, logic requires brain exercise with a little bit of ingenuity. A healthy mix of practice and actual problem-solving experience makes for a more logical person.

Emotional Intelligence. This is also known as EQ, and refers to one's personality and emotional patterns. Unlike creativity, however, these tests will purely be for personality assessment, and there is no right or wrong answer. These tests will help you determine, through a series of questions, the following, among others:

- Are you an introvert or an extrovert?

- Are you an optimist or a pessimist?

- How anxious or relaxed are you?

- How self-confident are you?

EQ tests are usually measured thru a point system, where answers will correspond to a particular score, and the numeric result will represent what your personality is between two extreme traits (i.e. optimism to pessimism). The results will then provide general advice as to how a person of your personality can improve to have better relations with people.

These tests help you understand yourself better so that you can start to improve certain parts of your personality. Remember, even if a person is smart, he or she can still benefit from improving control over their emotions.

Memory. As mentioned before, this refers to how well you can remember stuff. We practice memorization all the time (i.e. when we make a mental note of what to get at the grocery, etc.), but people rarely ever take time to exercise this ability. You'd be surprised at how much more short-term or long-term memory you can store if you practice.

Memory tests often require more time than others because they will ask you to look at a set of information or images for five minutes or so, before being faced with a question which usually requires you to recall what you just saw. For instance:

Study these five words: Red, Woman, Pencil, Dog, and House.

What is missing? House, Woman, Pencil, Dog, _____

This is a fairly simple example, but the point is that you will need to remember all five words to be able to know that the missing word from the set is Red. Of course, real memory tests tend to be more challenging.

Improving Your Aptitude

Again, the tests mentioned above are only designed to measure your current aptitude. But testing often and practicing on a daily basis can help you improve your logic, creativity, etc. If you want to know how you fare, consult an expert who can administer a test for you. While you can find most of these tests in books or online, an expert can help you assess the results better.

Chapter 3: Going Beyond the Test

Taking tests and knowing what to do to become better at certain aptitudes is only half the battle. The thing with IQ is that it has to be consistently trained to be sustained, if not improved. You also have to make adjustments within yourself, as well as the world around you if you want to see an increase in your intelligence.

Self-Motivation

Earlier in this guide, it was stressed that being aware that you can improve your IQ is essential to that end. This is because, much like all other self-development tracks, you're going to need motivation to be able to accomplish all the steps in between.

Remember that IQ is measured based on how you fare compared to others. Naturally, getting a higher IQ means working harder than everybody else. Going that extra mile means dedicating a bit of your precious time to go through exercises and perhaps changing some daily routines. They won't be easy, but they will be worth it.

- Motivation, for one thing, is needed in order to establish habits. Repetitive behavior can be monotonous and discouraging if you don't know what you're doing.

- It is also needed to stick to certain lifestyle changes. It's what will convince you that the changes are worth it.

- Lastly, it is needed in order to get back up in case some things don't work along the way. Much like the concept of IQ itself, improving intellect is not an exact science. You won't always reach your goals and you won't always get the expected results. Motivation is what will make you rethink your methods, or at least tell you to keep going even when you don't see any results yet.

Note that in all the aforementioned instances, a genuine realization that you can improve your IQ is very important.

Of course, there are many ways to get yourself pumped up to improve your IQ. Learning how to be motivated is the subject of a totally different guide in itself. The takeaway from all this is that you can very well increase your IQ if you just want to.

Environment

In contrast to motivation, which is an adjustment from within, your environment is also an important external factor to improving your IQ.

Studies have shown that the environment in which you thrive in factors greatly in your brain's development. Even without being exposed to scientific formula or statistics (which you don't even need to bother with), it's very easy to tell how your mind can be affected by your own surroundings. Too much noise, for example, can keep us from thinking straight. You might also find it hard to focus in a place that has poor lighting or ventilation.

Those examples, of course, are pretty obvious. There are more complex but subtle situations that most people don't know can affect their IQ.

Commonly Overlooked Habits and Environmental Factors

Here are a couple of factors that, as you may or may not have realized, contribute to your intellectual development:

- **Personal Time** Everybody needs alone time. You'd be surprised how a one hour of just being alone, doing whatever you want, can make such an impact to your otherwise dragging daily 8-hour work routine. Being at the right place at the right amount of time can be very gratifying. Consider it as something you can look forward to after attending to your responsibilities.

- **Adequate Sleep.** Getting that eight consecutive hours of shut-eye is not just about closing your eyes and waking up eight hours later. You need sleep in areas that are conducive for sleep. That means, ideally, a nice and comfortable sleeping area (i.e. your bedroom at home), and not your office desk with the lights out.

- **Mental Exercise.** This has been mentioned earlier, but it cannot be stressed enough. In the same way that you make your own schedule and place for studying, you should make time and space for those IQ improvement exercises.

> **Tip:** Take notes. Or make flashcards or memory aids. Try different techniques to figure what works best for you.

- **Proper Pacing.** This includes two things. First, don't work too much. Working more than you should only means you're getting less rest. Only do overtime when it's necessary, and do so sparingly. Second, do not cram. People who distribute their work load evenly will accomplish a task with less stress than a person who would accomplish the same task in the 11th hour. Plus, you run the risk of not doing the job right. Proper pacing guarantees that you can think about what you're doing so that your brain can work with what it's capable of and at the same time learn new things.

- **Physical Health.** Observe a proper diet and get enough exercise. It does wonders for the brain.

- **Being 'at-home' with your surroundings.** Be comfortable where you are (unless it's not your home). Some people like it organized, while others like an organized mess. Remember that only you can determine whether or not you're okay with your surroundings, so you have to do it yourself.

- **Attentiveness** The world around you tends to reveal more things than you think they can. Make it a point to be aware of your surroundings and the little things that happen no matter where you are. Is it raining? Did you remember to turn the lights off before leaving? Does everyone in your office seem okay? Not only is this a good exercise, it can also spare you from unnecessary consequences brought about by neglecting what seem to be minor details.

Indeed, there are more habits and details in your environment that you can manipulate to make your life more conducive for self-improvement, but the ones above will be enough to give you a jumpstart.

Chapter 4: Lifestyle Tips to Improve IQ

To review, IQ is never an end in itself. Its improvement is supposed to translate to how you are able to function in your daily life. This means being better able to solve problems, assess situations, accomplish tasks, and even relate with people better.

Make it a point to learn something new. If intelligence is the capacity to learn new things and practice is what improves that capacity, what better way is there to improve your IQ than to consistently try to learn new things? You may not necessarily learn things as well at first, but allowing your mind and body to be accustomed to a regular dose of new experiences can help greatly.

Share what you know with other people. First, this is a great avenue for developing emotional intelligence, which is also an important aspect of IQ. You get to connect with people and show them that you want to help. Secondly, being able to effectively convey your knowledge to others (especially to those who have poor comprehension skills), helps cement your own knowledge. It simplifies even the most complex of ideas, making it easier for you to master.

Minimize downtimes. Don't make "waiting" an excuse to not be productive. Always bring a book to read with you for those times when you're waiting for an appointment. Listen to music or audiobooks when you're on the bus to work, or talk and be friends with the people you're riding with. You'd be surprised how much of our day is spent getting from point A to point B.

Think of new ideas. Push your brain to its limits by trying to figure out stuff that others can't. Even the simple act of evaluating your own lifestyle and asking whether or not you're running things efficiently on your end can be quite the challenge. This is where all that logic and creativity comes in. Innovate!

Accept criticism. Nobody ever grows by being right all the time. Sometimes we have to listen to what others think we may have done wrong so we can improve. This is where your intrapersonal intelligence comes in. Instead of trying to waste time denying your mistakes, own up to them and just get on with resolving it. That does more than bickering with others about who to blame.

If you must, argue with evidence. Of course, don't be a pushover and just agree with everybody. If you think someone is wrong, prove it to him or her factually and allow that person to accept the mistake. After all, you would too, if you were wrong.

You'll realize that going back to past life situations with your newfound awareness for IQ improvement will help you handle future situations much better. In the end, it's not enough that you improve your IQ, but you should live a life with that improvement at play.

Chapter 5: Using Decision Making to Enhance Intelligence

So you've finally assessed your current standing with IQ tests, practiced a lot to improve them, and applied your new skills in real life. The next step is to go back to evaluating what you've done so far.

This doesn't mean you should take another IQ test. This time, view the situation with less "science" and more practicality in mind. Instead at looking at IQ in a general sense, examine a real life scenario or dilemma, particularly one you are facing at the moment.

This last chapter hopes to help you get better at decision making. There are three steps to objectively evaluate one's choices:

1. **Map and Plan.** Here, you are yet to make a decision. You will focus on assessing the current situation (i.e. what the problem or goal is), and then making preparations for the task. Remember to take into consideration all details. Naturally, your objective is to come up with something that will generate success.

2. **Take action.** This time, put your plan into action. Try to make sure that everything goes according to plan (although we all know that almost never happens).

3. **Assess the results.** This is the most important part. Given the outcome, how did you handle a particular task or situation? Were there things you didn't anticipate? What could you have done to prevent those mistakes? Most importantly, what were the things that were within your control that you could have done but didn't?

Oftentimes, we perform evaluations for group endeavors and for work, which is okay. But remember that even the smallest personal decisions can be evaluated as well. For instance, when you go out to buy groceries, ask yourself if you took the most efficient route to all the stops in your checklist. Or, if you were trying to bargain with someone, assess whether you could have done something to get a better deal.

At this point, you will realize that improving your IQ boils down to a cycle of repetitive action and evaluation. In truth, that's what consistent improvement in almost anything is all about.

Once you've learned to consistently improve your decision-making abilities, you've essentially put yourself in a situation where you are maximizing your own IQ while making room for growth.

Conclusion

There is no doubt that being smarter is always a good thing. It helps you land a better job, deal with people better, and be generally happier with yourself.

It's true that a lot of a person's cognitive development starts during the earlier stages of development, and as far as the early development years are concerned, you are either lucky or you're not. While you can never bring back those years, you can still do something about improving your intelligence starting today

Intelligence quotient is not just something that can be improved, but something that can be improved in countless ways through endless opportunities that are present throughout our lives.

Improving your IQ will always require practice. By being fully aware of this, you can keep an eye out for yourself. You can always make decisions and take action with both application and improvement in mind. Using your IQ in real life is the best kind of practice there is.

Don't be frustrated if you find others who are better than you in certain aspects of intellect. That is completely normal because we are all different. Look to what you are good at and work on that. It'll make you feel better about yourself. Remember that although improving your IQ has its perks, its purpose is to make you feel happier and contented with yourself. Work at staying optimistic and have fun while learning about yourself.

Read up on other guides and books and strive to learn continuously. There are publications dedicated specifically to improving intelligence and the different types of IQ tests. Furthermore, you can consult experts who can test you and advise you better based on your particular situation. Don't hesitate to reach out to others for help.

By improving your IQ, you basically make yourself a better person in so many ways, making yourself as well as others happy by coming up with new ideas and solving problems, may it be at work, with friends, or life in general.

Finally, I'd like to thank you for purchasing this book! If you enjoyed it or found it helpful, I'd greatly appreciate it if you'd take a moment to leave a review on Amazon. Thank you!

Made in the USA
Coppell, TX
23 February 2023